A GUIDE

TO THE

Twelve Tissue Remedies
of Biochemistry

THE CELL-SALTS, BIOCHEMIC OR
SCHUESSLER REMEDIES

BY

E. P. ANSHUTZ

PHILADELPHIA

BOERICKE AND TAFEL

1909

CONTENTS

194348

PREFACE

MEDICINE, in the immediate past, has been divided into three great divisions— Allopathic, Homœopathic and the Eclectic.

The Allopathic principle is "opposites;" if the heart beats too fast, give a depressant; if the bowels do not move, give a cathartic; if they move abnormally, give an astringent.

The Homœopathic principle is "similars" —the opposite to the Allopathic; to give a drug that will produce a similar tumultuous heart action to that experienced by the patient, or the loose or bound bowel, but in a smaller dose than will cause the abnormality in the normal human body.

The Eclectic principle is not very clearly defined. The word from which it is derived means "to pick out," "to choose from," and thus, in a sense, all physicians are Eclectics. Really Eclecticism is a development of folk medicine, the medicine of the herbalists,

from whom the world derives so many of our
best remedies, developed in the light of
modern drug science. It has no formulated
principle.

Many of the Allopathic division of late
years have come to adopt the designating
title "Scientific," the root of which is, "to
know," "knowing." While it is true that a
great store of facts have been discovered, the
knowledge of the cure of disease has not
made any marked advances, i. e., therapeutic
advances, save on the line of pure experi-
mentation.

To the three great divisions, has not the
time come to add a fourth, the Biochemic—
life chemistry?

Biochemistry is the discovery, or the
claimed discovery, that an abnormal state
of one, or more, of the twelve chemical substances ("the twelve tissue remedies") that
enter into the make-up of the cells of the
normal human body, occurs in disease, and
the administration of the disturbed cell-salt,
or chemical substance, at fault, in proper

dosage, is the one scientific therapeutic
measure that can correct it. These twelve
chemicals, or substances, it is claimed, con-
tain all that is, therapeutically speaking,
useful in curing disease. The brilliant suc-
cess that has attended their use lends color
to the claims made by Schuessler, but whether
they are all-sufficient therapeutically is a
question that must remain open. Their
value is so marked, however, that every
physician would do well to familiarize himself with them. The science of Biochemistry
claims that the undoubted drug cures of other
systems are effected by the fact that the
needed Biochemic salt was in the drug making the cure.

This book is but an orderly and more convenient arrangement of the material left by
Schuessler, and of accumulated experience
of others, laying no claim to much original
material. The skill of the physician must
be shown in tracing a given case of illness
to its disturbed salt; when that is done, the
remedy is apparent. It is really a system of

therapeutics that goes hand in hand with pathology. "Pure pathology," said Huxley, "is that branch of biology which defines the particular perturbation of cell-life." Thus it is that Pathology points out the disturbance in cell life that is as near the origin of disease as science has arrived and Biochemistry furnishes the only scientific key to a cure.

Science sees that the root of what man terms disease is a "perturbation of cell-life," but when it looks for the cause of that perturbation, it looks at a closed door, or it looks beyond physical science to the shadowy realms of metaphysics, which to the man of science is an unreal maze, a mental phantasmagoria in which man becomes lost, yet in it may lay the primal Cause of Disease.

PART I.

THE THEORY, DOSAGE, ETC., OF BIOCHEMISTRY.

DR. SCHUESSLER.

Dr. Wilhelm Heinrich Schuessler, the founder of the system of therapeutics known as Biochemistry, was born at Zwischenaln, Germany, on August 21, 1821, and died March 30, 1898, at his home in Oldenberg. He studied medicine and kindred sciences at Paris, Berlin, Geissen and Prague. He received his medical diploma at Geissen. He took up the study of Homœopathy while a practicing physician, and from this evolved Biochemistry. His one work on the subject is his *Abridged Therapy*, the 25th edition of which he saw through the press in the year of his death. An English translation of this 25th edition has been published.

The Theory of Biochemistry.

In his *Abridged Therapy*, Dr. Schuessler credits the germ of the idea to Dr. Moleschott's work, *The Cycle of Life*. He quotes from it to the effect that without a basis yielding gelatine there can be no true bone, nor true bone without bone-earth, nor cartilage without cartilage-salts, blood without iron, nor saliva without potassium chloride. The ashes of a human body form its earthy basis. It was from this idea that Schuessler developed his biochemic therapy. If anyone wishes to go into the theory fully, he should procure a copy of Schuessler's own work.

Twelve elements constitute the earthy, or mineral, basis of the cells of the body. A disturbance, deficiency or abnormality, of these salts causes disease. The administration of the needed salt reduced as nearly to an atom as possible, restores the body to health. "The biochemical method supplies the curative efforts of nature with the natural material lacking in the parts affected, *i. e.*, the inorganic salts."—*Schuessler*. The fact that, theoretically, the biochemic salts act curatively by supplying, in atomic form, a deficiency in the body, has caused some men who practice it, who are not physicians, to plead, when arraigned for practicing medicine without a license, that they were administering, not medicine, but "a food." Biochemistry, however, pertains to the practice of medicine in the highest sense.

BIOCHEMISTRY AND HOMŒOPATHY.

Owing to the fact that two of the great remedies of Homœopathy, *Silicea* and *Natrum muriaticum*, are used by both systems, it has been affirmed that they are essentially the same, especially as both use the infinitesimal, or atomic, dose, but this is an error. The uses of the biochemic remedy are discovered by chemistry alone, while the uses of the homœopathic remedies are ascertained by provings, by the pure drug effect, on the human being. As Schuessler puts it, Biochemistry acts "by means of homogeneous substances, while Homœopathy attains its curative ends in an indirect way, by means of heterogeneous substances." Indeed, some biochemists claim that it is the biochemic remedies, or salts, found in the various homœopathic remedies that act curatively. But as this book is concerned with the practical working of the biochemic remedies only, this subject will not be considered.

NECESSITY FOR THE INFINITESIMAL DOSE.

Schuessler writes: "The use of the small doses for the cure of diseases in the biochemical method is a chemico-physiological necessity." Again: "Every biochemical remedy must be thus attenuated, so that the functions of the healthy cells may not be disturbed, and yet the functional disturbances present may be equalized." Thus it will be seen that if one would get the benefits of biochemistry, he must get rid of the foolish old notion that if a little will act, a bigger dose will act quicker and better. As a matter of fact, the big, crude dose acts, as a rule, detrimentally.

DOSAGE.

Schuessler writes: "In my practice I generally use the 6th decimal trituration." This is noted on labels thus "6x." Two remedies, *Ferrum phos.* and *Calcarea fluorica*, he gave in the 12th decimal trituration, *i. e.*, the 12x. As to the frequency of the dose, he writes: "In acute cases, take every hour, or every two hours, a quantity of the trituration as large as a pea; in chronic cases, take as much three or four times a day, either dry or in a teaspoonful of water." A good general rule is to give five grains of the trituration, or its equivalent, five one-grain tablets, at a dose. The frequency must depend on the character of the disease.

How the Remedies are Prepared.

One part of a given salt in its crude form with nine parts of pure milk sugar is put into a power-driven mortar with four pestles, and triturated (ground) for four hours. This makes 1x. One part of this is again mixed with nine parts of milk sugar and triturated for two hours. This makes 2x. And so on to the 6x, 12x or 30x. These triturations may be administered in this "powder" form or in tablets, the latter being much more convenient. The tablets are made by moulding the trituration into that form of one grain each. From three to five tablets constitute a dose.

In the 6x or 12x there is not the slightest possibility of an overdose; twenty-five tablets can be as readily taken as five, but nothing special is to be gained by increasing the dose above five tablets, or an equivalent amount of the trituration. The action of a Biochemic remedy does not depend on a greater or less quantity, but on its fitness to the disturbed cells.

We have indicated the proper manner in which these remedies should be prepared, but there is another method by which the cost may be greatly reduced, namely, by making a 1x or 2x trituration and then simply mixing this with powdered milk sugar in the proper proportions without the expense of further trituration. But this method defeats the end sought by the comminuting process of trituration, which is to reduce the salt to atoms or molecules. Remedies prepared in this manner, while cheaper, are sometimes detrimental to Biochemistry, especially as there is no known means of distinguishing them from the true remedies. The reason for this is that, to be biochemically effective, the various salts must be reduced to the minutest possible atom or even ion, and this can be done most effectively by trituration with the sharp crystals of the sugar of milk very prolonged and very thorough.

The Selection of the Remedy.

In the accurate selection of the remedy lies the secret of the successful practice of Biochemistry. But the question in a given case of illness often arises: What salt, or even salts, are lacking in this special case? For instance, *Magnesia phos.* is contained in the blood corpuscles, brain, spinal marrow

nerves, bones and teeth, and a lack of it, or
a "disturbance," causes a great many and
varied forms of illness, evidenced by symptoms, objective and subjective; but other
salts are also contained in these parts.
Which is the remedy? The skill of the prescriber and his pathological and biochemical
knowledge must determine the remedy.

Some knowing ones have hit upon the very
obvious idea, to the ignorant, of mixing the
twelve remedies and thus simplifying the
practice of medicine. They argue that if a
cure is to be found in these twelve salts, give
them all at once and let the needed one do
the work. Very plausible, but it will not
work. The reason why it fails may be seen,
in a measure, by an illustration in the season-
ing of food. You select one or two condi-
ments for a given dish and the result is added
zest and gusto; but if you were to mix twelve
seasonings, each excellent in its place, the
result would be nauseating and rejected by
every epicure. Indeed, some of the salts
have diametrically opposite actions and

would neutralize each other. Schuessler is very severe on even the habitual alternation of only two remedies.

There is another view of this matter, which at best is quite obscure, namely, the fact that all the salts needed by the average human body are to be found in the food he eats. Iron, for instance, is found in several vegetables, yet if there is a "disturbance," as Schuessler terms it, calling for *Ferrum phos.*, the eating of those vegetables will not supply the lack, neither will the swallowing of iron filings do it, or taking a dose of crude phosphate of iron. Why this should be so, no one, probably, can answer; one skilled in biochemistry can see that *Ferrum phos.* 12x is needed and will effect a cure. Why iron in this form will cure while in the vegetable form, or as iron filings, or the crude phosphate, it will not, is an unsolved problem. In this suppositious case iron is the remedy and will effect a cure; the other eleven remedies will not cure the case and for some unexplained reason, neither will iron (*Ferrum phos.*) if mixed with the other eleven.

This peculiarity is still more strikingly illustrated in the action of that really powerful remedy, *Natrum muriaticum*, common table salt. We take this daily in very material doses with our food, yet in this form it will have no curative effect, even if needed in disease, but when indicated in abnormal states, it will act brilliantly if administered in the 12x or the 30x trituration. This incontestable fact also shows why it is that triturations which are made by mixing 1x of *Natrum mur.* with proportions of powdered milk sugar to make it the 12x, is detrimental to the practice of Biochemistry and a cruel wrong to invalids.

Man can state and realize certain facts, but very often in explaining them he can advance theories only. We explain the fact that a nut detached from the tree falls to the ground by the law of gravitation; when asked to explain gravitation, we must fall back on theories or be silent. This little

book is a presentation of the facts connected
with the Biochemic remedies. When it
comes to explaining their action, we can but
give theories and probably our theories are
neither better nor worse than those of other
men. It is as positively known that the
remedies act curatively as that the apple
falls to the ground, but *how* they act is at
best but a matter explainable by theories
only.

DURATION OF TREATMENT.

Relief may be almost instantaneous, as in the
case at times of the action of *Magnesia phos.*,
or it may require many months, as in cases
of nervous prostration, or paralysis, or some
of the lingering ills. And again, there are
cases that are beyond hope—incurable. As
long as there is perceptible improvement, the
treatment should be continued. When the
cure is effected, naturally the treatment
ceases. Beyond this generality, nothing
more definite can be said and the prescriber
must individualize each case. Each one is

a human being, but each human being is an entity, is himself, and there is no exact duplicate of him, or her, in the universe.

CONCLUSION.

Such, in brief, is an outline of the theory and practice of Biochemistry, a wonderfully efficient system of medicine for the relief and cure of human ills; one apparently simple yet most profoundly deep, for it deals with the cells and these are the primary of natural life, its beginning. It is a broad system that lends its aid in any method of therapeutics even where it is not used exclusively. It is a higher, perhaps the highest, form of chemistry and should be so regarded by the practitioner.

PART II.

MATERIA MEDICA.

The remedies of Biochemistry are variously known as "The Biochemic remedies, "The Twelve Tissue remedies," "The Schuessler remedies," "The Cell-salts," and possibly under other names in various localities. The following is a list of the remedies with customary abbreviations, though these latter are varied by different writers.

LIST OF REMEDIES WITH ABBREVIATIONS.

Calcarea fluorica, abbreviated *Calc. fluor.*

Calcarea phosphorica, abbreviated *Calc. phos.*

Calcarea sulphurica, abbreviated *Calc. sulph.*

Ferrum phosphoricum, abbreviated *Ferr. phos.*

Kali muriaticum, abbreviated *Kali mur.*

Kali phosphoricum, abbreviated *Kali phos.*

Kali sulphuricum, abbreviated *Kali sulph.*

Magnesia phosphorica, abbreviated *Magn. phos.*

Natrum muriaticum, abbreviated *Nat. mur.*

Natrum phosphoricum, abbreviated *Nat. phos.*

Natrum sulphuricum, abbreviated *Nat. sulph.*

Silicea, abbreviated *Sil.*

The nomenclature is that of the older German writers: *Calcarea fluorica* instead of *Calcium fluoride, Calcarea phosphorica* instead of *Calcium phosphate,* and so on.

The following is an outline of the general biochemic sphere of each drug. They are considered in alphabetical order.

Calcarea Fluorica.

(Calcium Fluoride.   Fluorspar.   Fluoride of Lime.)

This substance is found in the surface of the bones, enamel of the teeth, elastic fibers and the skin. A disturbance, or deficiency, in the molecules of this element is shown by lumpy, more or less hard, growths, horny growths, hard crusts with cracked hands, hardened mammary glands, or other parts. Osseous (bone) tumors from injuries, glandular swellings, cataract of the eye, gum boils, displacement or relaxed uterus, sagging abdomen, hardened exudations, hæmorrhoidal knots, varicose veins, and all diseases originating in the elastic fibre and bone covering, which is practically the same thing.

This remedy has proved curative in hard swellings on the bones, varicosis, varicose veins, whitlow, induration of testicles, suppuration, psoriasis, prolapsus uteri, internal piles, ozæna, eczema, knots, gout, defects in enamel, backache, corneal diseases, cataract of the eye, after-pains, bone bruises, catamenia, cough, etc., where the above general conditions prevail.

Calcarea Phosphorica.

(Phosphate of Calcium. Calcium Phosphate. Phosphate of Lime.)

This element, *Calcarea phosphorica*, is found in all cells, especially the bone cells, and, therefore, is necessary in the normal formation of new cells. Its absence, or disturbance, is shown in anæmic states; the weakened conditions following acute diseases; delayed formation of bones in the young; in the knitting of broken bones. Broadly speaking, it is the remedy for weaklings, the ill-nourished, the very young or old, for rickets, chlorosis, open fontanelles, pains accompanied by formication, numbness, cramps, cold sweat, convalescence, convulsions of teething children; stunted children, unhealthy obesity and for the many named and unnamed disease states that have their origin in, and are accompanied by, these general conditions. It has been called the Cell-wall Salt.

This salt has proved curative in cancer, catarrh, chlorosis, "green sickness," con-

sumption, rickets, curvature of the spine,
spinal weakness, open fontanelles, hydrocephalus, bow legs in children, scrofula,
albuminuria, anæmia, diseased state of bones,
broken bones, convalescence, convulsions,
emaciation, painful gums, dentition, headache, leucorrhœa, perspiration (excessive),
night-sweats, rheumatism, scrofulous ulcers,
clammy feet and hands, tonsils chronically
swollen, etc., when the above general conditions prevail.

Calcarea Sulphurica.

(Calcium Sulphate. Gypsum.)

In his later years, Schuessler was inclined
to omit this element from his list of "tissue
remedies," but the experience of physicians
familiar with it is against doing so, for while
its sphere is limited, it has been, as stated
by Schuessler, "successfully used in many
diseases," but he adds, "it does not enter
into the constant constitution of the organism," from which it might be inferred that
at times other remedies are temporarily use-

ful. *Calcarea sulphurica* is useful generally
in suppurations, of pus mixed with blood,
yellowish discharges from eyes, ears and nose,
of this character; boils, carbuncles, wounds,
expectoration, pimples, scabs, skin diseases,
sores, etc., etc. Yellow-colored suppuration,
or discharge, especially if tinged with blood,
a general swelling of soft parts threatening
pus, or an established vent oozing pus is a
call for *Calc. sulph.*

This salt has proved curative in ear diseases with matter streaked with blood, diarrhœa, boils, abscess, ulcers, carbuncles, fes-
tering sores or hurts, pimples, oozing scabs,
herpes, catarrh, suppurating glands, etc.,
when the above-named conditions prevail.

Ferrum Phosphoricum.
(Phosphate of Iron.)

This element is contained in the red blood
corpuscles and possesses "the property of
attracting oxygen" from the inhaled air.
Any disturbance in this salt is shown prim-
arily by relaxed muscular tissue with inflam-

mation, fever, heat, congestion, throbbing, swelling, with the innumerable succeeding ills. From its character *Ferrum phos.* is necessarily almost always the first remedy required in every attack of acute illness. It is the first in every *itis*, from bronchitis on through the list to tonsilitis. It is of the blood and is, as Rademacher said, a "universal remedy." It precedes and then the other remedies follow in acute ills. In long-seated ills, this remedy does not always hold true. It is also required in some anæmic conditions and those following loss of blood.

This remedy has proved curative in bronchitis, gastritis, laryngitis, mastitis, nephritis, pericarditis, periostitis, phlebitis, pleuritis, tonsilitis, as a primary remedy; abscess, rheumatism, hæmorrhages (red blood), boils, catarrhal colds, congestions, inflammations, coughs, croup, deafness, diarrhœa, earache, erysipelas, swellings, gonorrhœa, gum-boils, hæmorrhoids, nose-bleed, fever, headache, irritations, neuralgia, palpitation, pleurisy, quinsy, scarlet fever, sprains, wounds, etc.,

when the indications noted above are in
evidence.

Kali Muriaticum.

(Potassium Chloride. Kali Chloratum. Chloride
of Potash.)

This element, or salt, is chemically related
to fibrine. Like *Ferrum phosphoricum*, it is
found in almost all cells. "It will dissolve
white, or grayish-white, secretions of the
mucous membranes and plastic exudations," consequently it is called for in
catarrh, croupous, and all plastic exudations
of mucous membranes that are white or
grayish-white. It almost always follows
Ferrum phos. for the secondary conditions or
states succeeding inflammation. It is indicated in the white, or grayish, coated tongue,
in catarrh and diseases showing that peculiar
color of the exudations of the mucous membranes. It is especially useful in ear diseases,
catarrh of the middle ear and closed Eustachian tubes, cracking noises, deafness, etc.;
coughs, hoarseness and bronchial troubles;

sluggish liver, light-colored stools, lung diseases showing characteristic colored expectoration. Like *Ferrum phos.*, it is a useful inter-current remedy in many cases.

This salt has proved curative in abscess, adhesions, rheumatism, asthma, bronchitis, congestions, "colds," catarrh, coryza, constipation, cough, croup, deafness and ear diseases, chancre, diarrhœa, diphtheria, dropsy, dyspepsia, eruptions, exudations, excoriations, gastritis, gonorrhœa, hæmorrhage of dark, clotty blood (*Ferr. phos.*, bright blood), headache, hoarseness, whooping cough, indigestion, liver diseases, lung diseases, measles, meningitis, menstruation, mumps, orchitis, pimples, pleurisy, childbirth fever, fevers secondary stage, pains rheumatic, scarlet fever, sick headache, stomach derangements, swellings without fever, sycosis, throat diseases, quinsy, mucous ulcerations, vomiting dark blood, diseases of mucous membranes, etc., when the above-named conditions in general prevailed.

Kali Phosphoricum.

(Potassium Phosphate. Phosphate of Potash.)

This is the chief "nerve salt," and is found in the brain cells and nerve fluids, the blood plasma (white corpuscles) and the inter-cellular fluids. A disturbance, or deficiency, in the normal state of this physiological element shows, mentally, in despondency, anxiety, fearfulness, weak memory, mental decay, mental and physical break-down, neurasthenia, hypochondria, hysteria, insomnia, night terrors, irritability, insanity and paralysis; also in septic conditions, typhus, scurvy, ulcerations and physically degenerated states generally, foul states and discharges, gangrene, blackish, thin blood, putrid conditions. Decay generally, ment a and physical.

This salt has proved curative in nervousness, neurasthenia, anxiety, gloom, morbidness, depression, brain-fag, loss of memory, sleeplessness, delirium tremens, horrors, dread, epileptic fits, hypochondria, hysterics, insanity, nervous diseases, paralysis, night terrors, palpitations, exhaustion, softening of the brain, sexual weakness, septic states, typhus, typhoid, gangrene, black small-pox, appendicitis, amenorrhœa, asthma, blood blackish but thin, Bright's disease, cancer, chancre, cholera, croup, cystitis, deafness, diabetes, diarrhœa, dysentery, diphtheria, eczema, epilepsy, nose-bleed, headache, hæmorrhage blackish and thin, rickets, putrid sore throat, sciatica, sepsis, spinal diseases, suppurations foul, etc. when the above named conditions are present.

Kali Sulphuricum.
. (Potassium Sulphate.   Sulphate of Potash.)

Like *Ferrum phos.* this salt aids in effecting the transfer of the inhaled oxygen. It is found in nearly all the cells containing iron. Any abnormality or deficiency is shown physically, by a sensation of weariness and heaviness, with chilliness, headache, pains in the limbs, or vertigo. The malaise is worse when patient is confined in rooms, and from warmth and is better in the open, or in cool, air. It is called for in desquamation following diseases like scarlet fever, measles, etc., scaly skin, yellowish eczemas, dandruff, etc. It is indicated in conditions of the mucuous membranes, exudations, discharges, secretions and tongue, when the color is distinctly yellow or yellowish.

Has been successfully prescribed in bronchitis, catarrh, albuminuria, diphtheria, dropsy, vertigo, asthma, ear diseases, whooping cough, syphilis, tuberculosis, stomach catarrh, anæmia, rheumatism, ophthalmia, lupus, eczema, skin diseases, measles, menstruation, leucorrhœa, indigestion, diarrhœa, disorders of the eye, etc., when general characteristics and general indications as above are shown.

Magnesia Phosphorica.
(Phosphate of Magnesia.)

This salt, or element, is found in the blood-cells, muscles, brain, spinal marrow, nerves, bones and teeth. A disturbance, or deficiency, of it is shown in sharp pain, cramps and paralysis. The patient is better from warmth and gentle but firm pressure, but worse from a light touch or jar. Its diseases take the form of neuralgia, toothache, faceache, convulsions, fits, cramps, colics, spasms, palsy, trembling, twitching, hiccough, tetanus and similar conditions. Its pains are likely to be of a lightning like character. It is the opposite to the other great nerve remedy, *Kali phos.*, which betokens degeneration and decay, while *Magn. phos.* is generally evidenced by acute pain.

It has been successfully prescribed in neuralgia, sharp pain, cramps, angina pectoris, chorea, St. Vitus's dance, epilepsy, colic, convulsions, cough, whooping cough, diplopia, dysentery, faceache, toothache, labor pains, menstrual colic, colic, ovarian neuralgia, puerperal convulsions, laryngismus stridulus, palsy, spasmodic retention of urine, spasmodic conditions, stricture, squinting, teething, tetanus, trembling, etc., when general conditions indicate a deficiency of this salt, as noted above.

Natrum Muriaticum.

(Sodium Chloride. Common Table Salt.)

When there is a disturbance or deficiency of the molecules of this salt, which attracts water, the water which should go to the cells remains in the intercellular fluids and hydræmic states ensue, that may be shown in various ways; patients tire easily, are chilly, have cold extremities and often an abnormal craving for salt. Malnutrition, emaciation, secretions of clear, watery fluid, tears, or clear, watery mucus. Clean, frothy tongue. Watery blisters and blebs. Coryza, clear, watery or frothy; watery eyes and much sneezing. Waterbrash, heart-burn. Intermittent fever. Dropsy. Puffiness. Mucous membranes may be abnormally dry.

This remedy has proved curative in intermittent fever, coryza, chlorosis, "whites," leucorrhœa, leukæmia, chlorosis, adynamic states, aphthæ, asthma, bronchitis, catarrhs, coryza, coughs, effusions, hay-fever, headache, diarrhœa, watery blisters, constipation,

deafness, water brash, house-maid's knee,
hydrocele, neuralgia, pleurisy, rheumatism,
sycosis, etc., when the general conditions of
Natrum mur. are in evidence, as noted above.

Natrum Phosphoricum.
(Sodium Phosphate. Phosphate of Soda.)

This salt is found in the cells of the blood,
muscles, nerves, brain and intercellular fluids.
A departure from the normal in it is characterized generally by "sourness," an excess of
lactic acid; sour eructations, or vomiting, or
diarrhœa in infants or adults; acidity; uric
acid conditions, rheumatism of the joints,
gout. Discharges from the eyes, or sores,
are yellow and creamy; tongue yellow;
throat and tonsils are covered with yellow,
creamy coating. Dyspepsia, sour or acid.
Perspiration smells sour. An anti-acid.

This salt has proved to be curative in indigestion, acid stomach, flatulence, sour ris-
ings, nausea, vomiting, worms, loss of appe-
tite, diarrhœa, headache, heart-burn, intermittent fever, morning sickness, scabs yellow,

sores, etc., when the general conditions noted above are present.

Natrum Sulphuricum.

(Sodium Sulphate. Sulphate of Soda. Glauber Salts.)

The action of this salt is the reverse of that of *Natrum mur.*, whose action is to attract the needed water to the cells; *Natrum sulph.*, on the contrary, eliminates waste matter and water. When there is a disturbance in its action, it is shown by a tongue of yellowish green or a dirty brown; discharges the same general color; biliousness; bitter taste; vomiting, of bitter water or bile; bilious diarrhœa; bilious fever; "yellow fever;" soreness in the region of the liver; jaundice. Worse from wet weather, dampness, moisture, etc. Better in dry and warm surroundings.

This remedy has proved curative in biliousness, bilious colic, jaundice, liver disorders, bilious headache, sick headache, vomiting bile, eructations bitter, diabetes, diarrhœa, dropsy, intermittent fever, yellow

fever, dengue, swamp fever, coast fever,
vertigo, etc., when the above-mentioned
general conditions are shown.

Silicea.

(Silica. Silex. Acidium Silicicum. Flint.)

Silicea is·a constituent of the connective
tissues, the skin, hair and nails. Disturbances in this element are closely connected
with pus formation and fistulous openings; the
skin is unhealthy and easily suppurates; the
nails are dry and brittle and the hair falls out
and is dry; perspiration is offensive, the feet
emit a bad smell, and so do the arm-pits;
atrophy of parts. Internal chilliness, sensi-
tive to cold air, wants to wrap up warm. In
headache must wrap up the head warm.

This element has proved curative in fistula,
boils, carbuncles, gouty deposits, rheumatism, hip joint disease, running sores, indurations, perspiration of the feet, pus secretions
of a rather watery nature, suppurations, styes,
ulcers, whitlows, chilblains, headache, etc.,

where the above general conditions are present.

NOTA BENE.

Names of diseases are given for convenience only. The seat of the disease lies in the aberrance of some one or more of the twelve tissue constituents, which may be manifested in different forms, according to the constitution of the individual affected. Pathology, and a true reading of the meaning of the signs, or symptoms, of the disease, must be the guide to a true biochemic prescription. Diseases bearing the same name vary as greatly as do the members of a family, and something more than correct nosology, classification, is needed for a curative prescription.

PART III.

THERAPEUTICS OF THE BIOCHEMIC REMEDIES.

Under this section is collected the general experience of physicians in treating, with the Tissue Remedies, the various diseases that here follow each other in alphabetical order. While this will serve as a general guide, it must be always borne in mind that back of the name lies the cause of the disease, *i. e.*, the deficient, or disturbed salt, in the organism. This must be constantly kept in view. The tissue remedy that may prove curative in one case of disease correctly diagnosed under a given name may not be the remedy for another case correctly diagnosed under the same name. Always keep the general character of each remedy in mind.

Abscess.—An abscess is a circumscribed cavity containing pus in any part. The chief remedy is *Silicea.* If in connection with hard, swollen glands, *Calc. fluor.* If especially foul, *Kali phos.* The color and character of the discharge may indicate some other remedy. *Ferrum phos.* is indicated at the beginning if there be heat.

Acidity.—For acid or sour stomach, acid or sour vomiting, eructations, or diarrhœa, *Natrum phos.*

Acne.—An inflammation of the sebaceous glands, accompanied by pustules; *Silicea.*

Addison's Disease.—A disease characterized by bronzed skin, anæmia, feeble heart and debility. *Natrum sulph.* is the remedy. An occasional dose of *Natrum mur.* as intercurrent remedy may be needed in some cases.

Albuminuria.—Latin, "white;" white of an egg. A condition in which the urine contains albumen. The constitutional symptoms must decide whether the remedy be *Kali sulph., Calc. phos., Kali phos.* or *Natrum mur.* (See remedies biochemically considered.) If the condition follows scarlet fever, the remedy is *Kali sulph.*

Alopecia.—From the Greek for "fox," because that animal sometimes sheds his hair in spots. Falling of the hair in spots. *Silicea* morning and *Natrum mur.* in the evening has proved beneficial. The same remedies are for the falling of the hair when not in spots. It is symptomatic of other disease.

Angina pectoris.—An intense, neuralgic pain about the heart, with difficult breathing and sense of suffocation, *Magnesia phos.*

Aphonia.—Loss of voice. If from illness, "colds," *Kali mur.* If from using it too much, public speaking, singing, shouting, etc., *Ferrum phos.*

Aphthæ.—Sore mouth, eruptions, thrush. If color be white or grayish, *Kali mur.;* if yellow, *Natrum phos.*

Apoplexy.—From the Greek "to strike," hence, our word "a stroke." The remedies are *Ferrum phos.* immediately, followed by *Silicea.* To guard against a possible stroke, take occasional doses of *Calc. phos.*, about twice a week.

Appendicitis.—An inflammation of the vermiform appendix of the cæcum; perityphlitis. The chief remedy is *Kali mur.* For intense pain, *Magn. phos.* Constitutional, *Silicea.* Fever, *Ferrum phos.*

Asthma.—From the Greek "to blow." Great difficulty in breathing, at intervals, with wheezing and sense of constriction; sometimes with cough and expectoration. In this disease, the symptoms and expectoration must determine the remedy. If nervous, *Kali phos.* Greenish-yellow mucus coughed up, *Natrum sulph.* Frothy, clear, or watery, *Natrum mur.* Lumpy, hard expectoration, *Calc. fluor.* Heart involved, cardiac asthma, *Kali mur.* Dry, painful, spasmodic, *Magnesia phos.* A course of *Natrum sulph.* is a good constitutional treatment to eradicate this disease.

Ataxia.—From the Greek "out of order." Want of co-ordination in muscular movements; a degeneration of the spinal cord. Locomotor ataxia, tabes dorsalis. Its final stage is bed-ridden paralysis. The remedy is *Kali phos.* A complete cure can hardly be expected. *Kali mur.* and *Silicea* may be substituted every four or five days for a day or two. *Magn. phos.* is indicated if there is sharp pain.

Atrophy.—From the Greek, "deprived of nourishment." Defective nutrition of parts. Marasmus. A condition that may occur in almost any part of the body. The chief remedy is *Kali phos.* In children, marasmus, delayed dentition, shrunken, *Calc. phos.* As an intercurrent remedy, *Silicea* will be found useful. Rapid emaciation of throat or neck, *Natrum mur.* In bottle-fed babies, *Natrum phos.* In infants with large head but wasted body, *Silicea.*

Backache.—It may come from a variety of causes, and, therefore, require different remedies. The symptoms must guide when cause is unknown. If from a cold, lumbago, rheumatic, *Ferrum phos.*, followed by *Kali mur.;* from nervous exhaustion, physical breakdown, *Kali phos.* Better in open, cool air, *Kali sulph.;* better from warmth, *Silicea.* Aching, tired, as if bruised, relief by mild exercise, *Calc. fluor.* Lightning-like, darting pain, *Magn. phos.*

Bed Sores.—Internally *Kali phos.* Also dissolve some of the trituration, or make a thin paste of it and apply externally.

Bed Wetting.—Enuresis. This annoying complaint will often yield to *Natrum sulph.;* if caused by worms, *Natrum phos.;* in feverish cases, *Ferrum phos.* Other remedies may be constitutionally indicated.

Biliousness.—The word is from the Latin *bilis,* "the bile." This disease is not a very well-defined complaint and is used popularly to describe a disturbed digestion, which may be accompanied by constipation, malaise, coated tongue or vomiting; bilious fever. *Natrum sulph.* is the chief remedy, indeed, the only one when the affection is purely bilious, *i. e.,* caused by the bile.

Bladder.—The many ills that affect the bladder and the remedies will be briefly summarized. *Cystitis,* inflammation of the bladder, with retention of urine or inability to retain it; *Ferrum phos.* Stone in bladder, *Calc. phos.* and *Silicea,* alternately.

Paralysis of the bladder, *Natrum sulph.* *Kali phos.* may also be called for.

Bones.—Necrosis, caries (dead) of bones, *Silicea;* also *Calc. phos.* may be required. Bone bruises, exostoses (tumors), hard growths, nodes, *Calc. fluor.* Diseases from mercury, *Calc. fluor.* Periostitis, inflammation of the surface of the bones, *Ferrum phos.* Broken bones do not knit readily, *Calc. phos.* Atrophy, *Kali phos.* Fistula, discharging splinters of bone, *Silicea* or *Calc. fluor.* The chief remedies in all diseases involving the bones are *Silicea, Calcarea fluorica* and *Calcarea phosphorica.* Hip joint disease, which is due to scrofula, requires *Natrum phos.* or *Silicea.* While other remedies may be called for in diseases of the bones, *Silicea* may be regarded as the primary one.

Brain Fag.—Men break down, cry like children, are unable to continue business, nervous prostration, neurasthenia, *Kali phos.* If this condition follows illness, weakness or night sweats, *Calc. phos.*

Bronchial Affections.—Bronchitis, inflammation of the lining of the bronchial
tubes, first remedy, *Ferrum phos.;* when this
is followed by coughing up of white or grayish mucus, second stage, *Kali mur.* Bronchial catarrh, tongue yellow, feels better in
the cool, open air, *Kali sulph.*

Callosities.—One with tendency to cal-
lous growths may need *Calc. fluor.*

Cancer.—From the Latin word for
"crab." A malignant growth that, crab-
like, fastens on one part. The knife of the
surgeon will give relief, and the operation is
often needed to save life, but it will not re-
move the constitutional cause, and generally
the cancerous growth returns, demonstrating
that medicines are needed to free the constitution from the cause of the ailment. The
guide to the remedy can only be general.
Hard, or firm, growths indicate *Calc. fluor* as
the constitutional salt needed. Yellow, thin
discharges, skin cancer, *Kali sulph.* The
cancer of the scrofulous, *Calc. phos.* For
those who are better in the cool, or cold, open

air, *Kali sulph.* For those who are generally chilly and seek the warmth, *Silicea.* It is generally held that cancers are incurable by medicine, but their growth may be modified, or arrested, by the right medicine. Some physicians maintain that cure is possible in many cases.

Carbuncle.—From the Latin, "a live coal." A deep-seated, suppurative inflammation. The remedy is *Calc. fluor.*, followed by, or in alternation with, *Kali phos. Ferrum phos.* is required for the fever.

Catarrh.—From the Greek, "running down," "to flow." This term is used rather indiscriminately, to denote everything from a bad "cold" to a chronic state. Ordinary "bad cold," *Ferrum phos.;* "sneezing cold," watery, *Nat. mur.;* dry coryza, *Kali mur.;* green discharge, *Nat. sulph.;* foul ozæna, *Nat. phos.* or *Magn. phos. Silicea* is also required in some cases of ozæna.

Chicken-pox. — Varicella. A disease characterized by vesicles over the body, which after a few days open and scab, but rarely pit, as is the case with small-pox. *Ferrum phos.* is the remedy for the first stage; *Kali mur.* when vesicles open, and *Kali sulph.* when desquamation sets in. If there are any lingering effects of the disease, after all appearances of disease have vanished, give *Silicea.*

Chilblains.—*Natrum sulph.* If suppurating, *Kali phos.*

Cholera.—From the Greek for "bile," hence, choler. Asiatic cholera is characterized by diarrhœa, rice-water discharges, cramps, vomiting and terminates in livid coldness, unless the reaction sets in. *Natrum sulph.* is the chief remedy, with *Magn. phos.* for the cramp stage, if it sets in. (See also "Colic" and "Diarrhœa.")

Chorea.—From the Greek, "a dance," hence "St. Vitus's dance." A disease characterized by involuntary, spasmodic or jerky motions of the limbs or body. *Magn. phos.* is the chief remedy, though *Calc. phos.* may be needed in some cases.

Colds.—To abort an ordinary cold in the

head, give *Ferrum phos.* 3x, every fifteen minutes for an hour or two. If cold begins with unusual frequency of sneezing and water running from nose or eyes, *Nat. mur.*, 12x.

Colic.—"The colon," originally a pain in the colon, but now used more generally. Flatulent, *Magn. phos.;* in umbilical region, *Magn. phos.;* with acidity, *Natrum phos.;* bilious, *Natrum sulph.;* menstrual, *Ferrum phos.;* painters' or "lead colic," *Nat. sulph.;* "gall-stone colic," *Magn. phos.*

Concussions of the Brain.—*Kali phos.* Inflammation, *Ferrum phos.*

Constipation.—From the Latin, "to crowd together." Fecal retention, owing to diminished action of the muscular coat of the intestines. It is not so much a disease *per se* as a symptom of disease. Constipation with white or grayish tongue, light-colored fæces, *Kali mur.* With occasional diarrhœa, in young children, "an admirable laxative," *Natrum phos.* With anæmia, pale, flushes of heat, palpitation, piles, *Ferrum* *phos.* Rectum seems to have lost its power, sweaty feet, or hands, paralytic, *Silicea.* Constipation of fleshy persons, drowsiness, water-brash, watery eyes, intestines seem dried up, *Natrum mur.;* the 30th will act better than the 12x.

Coryza.—See "colds."

Cough.—A cough is but a symptom of a disease or of the presence of some foreign body, and must be treated symptomatically. Short, spasmodic, painful, tickling, hard or dry, *Ferrum phos.*, followed well by *Kali mur.* Croupy, gray or white tongue, grayish or white expectoration, "stomach cough," *Kali mur.* Expectoration salty, fetid, *Kali phos.* Yellow, stringy expectoration, worse in warm room, *Kali sulph.* Whooping cough, *Magn. phos.* Clear, watery, glairy expectoration, *Natrum mur.* Yellowish, green, chest sore, *Natrum sulph.* Better from warmth, night coughs, night sweats, *Silicea.*

Cramps.—A sudden, involuntary muscular contraction, more or less painful, of any part. *Magn. phos.* is the principal remedy.

In the calves of the legs, *Calc. phos.* In cholera, *Nat. sulph.*

Croup.—An inflammation of the mucous membrane of the larynx which may include trachea, with fibrinous exudations, or membrane, causing a peculiar cough and a saw-like breathing. The remedy is *Kali mur.*, with *Ferrum phos.* as an alternate for the fever. Should the case grow steadily worse give *Calc. phos.*

Debility.—Nervous or sexual, *Kali phos.*

Delirium Tremens.—The majority of cases will clear up under *Natrum mur.* or *Kali phos.*

Dentition.—Teething. The chief remedy for weakly children is *Calc. phos.* Convulsions will call for *Magn. phos.* Very feverish *Ferrum phos.* Sweat, abnormal, *Silicea.* If the bones of the infant seem abnormal, *Calc. fluor.*

Diabetes.—From the Greek "to pass through." Excessive urination. Diabetes mellitus and diabetes insipidus. The difference is that the urine in the first-named is charged with sugar, the latter not. The disease is characterized by the passing of large quantities of urine, thirst and emaciation. The remedy is *Natrum sulph.* If there be, or develop, nervous prostration, *Kali phos.* may be required.

Diarrhœa.—From the Greek, "to flow through." A disease that may arise from a variety of causes and thus may require different remedies. The symptoms and cause must guide to the remedies. If evacuations are watery and mucous, *Natrum mur.;* carrion-like, *Kali phos.;* bilious, *Natrum sulph.;* undigested, *Ferrum phos.;* with griping colic, *Magn. phos.* Cholera, *Natrum sulph.* Caused by fatty food, pastry, etc., *Kali mur.;* by acidity, *Natrum phos.;* by damp weather, *Natrum sulph.* In scrofulous, rachitic children, *Calc. phos.* Bloody, red blood, *Ferr. phos.;* dark and thin, *Kali phos.;* dark and clotted, *Kali mur.*

Diphtheria.—From the Greek, "a membrane." A disease characterized by profound vital depression and the formation of

a grayish-white membrane in the throat;
ulceration, gangrene or paralysis may follow.
For typical diphtheria, *Kali mur.* is the remedy. With swelling and a white exudation,
which may cover the uvula, *Calc. phos.* If
exudation is markedly yellow, *Natrum phos.*
When the breath becomes putrid and the
throat gangrenous or the case paralytic, *Kali
phos.* Begin all cases with *Kali mur.*

Dropsy.—From the Greek, "water."
Hydrops. Ascites. Anasarca. Ascites ap-
plies to dropsy of the lower belly. Anasarca,
means general dropsy. The various phases
of this complaint generally arise from other
diseases. The best general remedy is *Natrum sulph.* Swelling of lower limbs, *Kali
mur.* Following scarlet fever or diphtheria,
Natrum mur. Hydrops genu, of the knee,
Calc. phos. Hydrocele, dropsy of scrotum,
testicles, *Calc. fluor.* Another remedy to be
considered is *Natrum mur.*

Dysentery.—A disease often called
Bloody Flux. An inflammation of the
mucous lining of the large intestine, charac-

terized by bloody evacuations and painful straining. *Ferrum phos.* and *Kali mur.*, in alternation, are the remedies. If the evacuations are very ill-smelling, *Kali phos.* may be substituted for *Kali mur.*

Dysmenorrhœa.—From the Greek words "difficult," or "painful," "month" and "flow." Difficult or painful menstruation. With congestion and fever, blood, *Ferrum phos.;* where this recurs monthly, this remedy will act as a prophylactic, if taken a few days before. Backache, vertigo, headache, sexual excitement, *Calcarea phos.* Menstrual colic, *Magn. phos.* If fever blisters be present, *Natrum mur.* Great coldness, *Silicea.* Acid, acrid, *Natrum sulph.* Long-standing, chronic cases have been cured by a prolonged course of *Kali phos.*

Dyspepsia.—From the Greek, "ill" and "to cook." Indigestion may arise from various causes. Hot, with fever, pain, *Ferrum phos.* Heavy feeling in the stomach, white tongue; from fat or rich food, *Kali mur.* Feeling of a load on the stomach, yellow

tongue, *Kali sulph*. Spasmodic cramps and
pains, *Magn. phos*. Water-brash, clean
tongue, *Natrum mur*. Acid, sour, *Natrum
phos*. Bilious, bitter, *Natrum sulph*. Excessive accumulation of gas, *Calc. phos*. In
long-standing cases, *Silicea* has often proved
curative.

Ear Diseases.—Diseases of the ear often,
but not always, spring from some constitutional cause. A cure must be sought in the
treatment of these ills. Noises in the ears,
and difficulty in hearing, *Ferrum phos*. Discharge of pus, *Silicea* or *Natrum phos*. Thin,
yellow discharge, *Kali sulph*. Catarrh of
Eustachian tubes, causing hardness of hearing, *Kali mur*. Hardened wax, *Calc. fluor*.
Ear troubles of nervous origin, *Magn. phos*.
Pulsations in the ear, *Ferrum phos*. Inflammation of the ear, earache from cold or wet,
Ferrum phos. Cracking noises in the ear
when gulping, *Kali mur*. Chronic catarrh of
middle ear, *Kali mur*. Foul-smelling, dirty
discharges from the ear, otorrhœa, *Kali
phos*. Suppuration and discharges from

scarlet fever, *Kali phos.* As a general remedy for the ear, *Kali mur.* stands first.

Eczema.—See Skin Diseases.

Epilepsy.—From the Greek, "to seize," "to attack." A nervous disease suddenly attacking one, causing convulsive movements and sometimes unconsciousness. "Fits." The principal remedy is *Kali mur.*, though others may be needed in connection with it, as *Ferrum phos.*, where there is a rush of blood to the head; *Magn. phos.*, if disease is result of vicious habits; fits come on at night, *Silicea.* In some cases of menstrual origin, *Calc. phos.* or *Kali phos.* Constitutional conditions must largely guide in the selection of the remedy.

Epistaxis.—From the Greek, "to drop." Nose-bleed. If blood is bright red, *Ferrum phos.;* if dark and coagulating, *Kali mur.;* if dark but thin, *Kali phos.*

Erysipelas—From the Greek, "red" and "hide," or "skin." "St. Anthony's fire." An inflammation of the skin in parts, characterized by fever and swelling. Vesicles, or eruptions frequently appear on the inflamed surface. The smooth, red, shiny, puffy erysipelas, *Natrum sulph.;* if skin breaks, infiltration, *Natrum phos.;* for high fever, *Ferrum phos.*

Exophthalmic Goitre. — From the Greek, "the eye," and "bulging," combined with the Latin for "the throat." The disease, an anæmic condition accompanied by a protrusion or bulging of the eyes in connection with the thyroid gland, is also known as "Basedow's Disease," "Graves's Disease," "Stokes's Disease" and other names. Schuessler tersely gives the treatment for "Goitre," "*Magnesia phosphorica.*" In connection, however, with this remedy, *Natrum mur.* may be useful. See also "Glandular Affections."

Eyes.—Blepharitis, inflammation of the eyelids, *Kali mur.* Styes, *Silicea.* Inflammation of eyes, *Ferrum phos.* Grayish-white secretions, *Kali mur.;* yellow, *Kali sulph.;* yellowish-green, *Natrum sulph.* Watery eyes, *Natrum mur.* Ulcers on the cornea, *Kali mur.* Cataract, *Calc. fluor.* "Pink eye,"

Ferrum phos. Burning eye-balls, *Magn. phos.* Weak sight, dread of light, watering eyes, *Natrum mur.* Fiery sparks, *Magn. phos.* Ulcers, pus, *Silicea.* Lids glued together in the morning, *Natrum phos.* Squinting, spasmodic, *Magn. phos.* Sensation of sand, *Kali phos.* Drooping eyelids, *Magn. phos.* Weak vision from over-use of the eyes, *Calc. fluor.*

Fistula.—Latin, a "pipe" or "reed." An opening into the flesh from which there is constant oozing or flowing. The chief remedy is *Silicea,* though some cases may require *Calc. phos.*

Flatulence.—*Mag. phos.* or *Natrum phos.*

Gall-Stones.—For the pain, *Magn. phos.* To prevent formation, *Calc. phos.*

Gastric Derangements.—Inflammation of the stomach, violent pain, *Ferrum phos.;* pain in stomach from eating, *Ferrum phos.* Cramps, *Magn. phos.* Water-brash, *Natrum mur.* Feeling of fullness, pressure, load in pit of stomach, tongue yellow, *Kali phos.* Umbilical colic, colic generally, *Magn. phos.*

Acid stomach, *Natrum phos.* Heart-burn,
ills from fat, *Natrum phos.* Bitter, bilious,
vomiting, *Natrum sulph.* Ulceration of stomach, *Kali phos.* Painters' colic, *Natrum phos.*
Enlargement of stomach, *Kali phos.* Dyspepsia that is temporarily relieved by eating,
Calc. phos. Neuralgia of stomach, *Magn.
phos.* Catarrh of the stomach, *Kali sulph.*
Acid things disagree, *Natrum mur.* Chronic
dyspepsia, *Natrum sulph., Silicea* or *Kali phos.*
(See also "Dyspepsia.")

Glandular Affections. — When the
swollen gland is firm or hard, *Calc. fluor.*
For fever, *Ferrum phos.* Scrofulous glands,
Calc. phos. Glands of throat swollen, *Kali.
mur.* Glands affected by vaccination, *Silicea.*
Suppurating glands, *Silicea.*

Gleet.—A chronic discharge from any
mucous membrane, usually applied to the
urethra; chronic gonorrhœa. The remedy is
Natrum mur.; some cases have been cured
by *Silicea* continued for months.

Goitre.—*Magn. phos.* (See Exophthalmic
Goitre.)

Gonorrhœa.—From the Greek, "semen" and "to flow." A discharge of mucus or pus from the urethra. The disease, unless properly treated, may infect almost any part of the body. Great care should be exercised to keep any of the discharge from coming in contact with the eye, else blindness may result. The chief remedy is *Natrum phos.*, according to Schuessler, though later practitioners of biochemistry have come to rely more on *Kali mur.* The two in alternation may be best. Where the case is of long standing and the discharge fetid, the remedy is *Silicea;* if clear, watery or slimy, *Natrum mur.;* should the discharge be bloody, *Kali phos.*

Gout.—From the Latin, "a drop," from an old but mistaken idea of the disease. Gout is an excess of uric acid, causing a deposit of urates (of sodium) in and around the joints. The term is usually applied to the disease in the great toe, but it may assail other parts. *Natrum phos.* dissolves, or counteracts, the uric acid, while the urates require *Silicea*. (See also "Rheumatism.")

Grippe.—(See "Influenza.")

Hæmorrhage.—Bleeding from any part. Of bright red blood, *Ferrum phos.;* of dark clotted or thick blood, *Kali mur.;* blackish, thin, putrid, like coffee grounds, *Kali phos.*

Hæmorrhoids.—Piles. The remedy is *Calc. fluor.* If there is much bleeding, alternate *Ferrum phos.;* or if very painful, *Magn. phos.;* if there be much mucus, *Natrum mur.;* when piles protrude or are incarcerated, suppurate, *Silicea.*

Hair.—When the hair falls out, *Silicea* and *Natrum mur.* may arrest the process, in some cases; one dose of each a day.

Hay Fever.—Hay asthma, summer colds, characterized by much sneezing, running of nose and eyes, headache, etc. *Natrum mur.* is the chief remedy if sneezing be a predominant symptom. If there be no sneezing, *Natrum sulph.* If there be fever, *Ferrum phos.*

Headache.—The treatment of this symptom, which accompanies so many ills, must depend largely on the cause of the headache. In general, *Ferrum phos.* will be found the remedy for headaches from cold, sunheat, throbbing, fever, congestion. *Kali phos.*, headache of students; nervous, irritable, weary, despondent. *Kali sulph.*, originating in warm or crowded rooms. *Magn. phos.*, neuralgic or rheumatic, excruciating pain. *Natrum sulph.*, sick headache, bilious headache, nausea, vomiting. *Natrum phos.*, frontal headache, or on top of the head. *Calc. phos.*, headaches characterized by a feeling of coldness. *Calc. sulph.*, headache characterized by vertigo. *Silicea*, for headache in cases where patient requires the head to be wrapped up warm.

Heart.—Sharp pain, *Magn. phos.* Enlarged heart and hardening arteries, *Calc. fluor.* Palpitation, *Ferrum phos.* Heart troubles following illness, *Kali mur.* Weak heart, patient nervous, depressed, short of breath on least exertion, *Kali phos.*

Hiccough.—*Magn. phos.*

Hip Disease.—Inflammation, *Ferrum phos.* Suppuration, *Silicea* or *Calc. phos.*

Hoarseness.—From cold, wet, or exertion of the voice, *Ferrum phos.* Long continued loss of voice, *Kali mur.*

Hydrocele.—From the Greek, "water" and "tumor." Dropsy or swelling of scrotum or covering of testicles. The chief remedy is *Natrum mur.*, which not clearing up the case, alternate with *Calc. fluor.*

Hydrocephalus. — From the Greek, "water" and "head." Dropsy of the head or brain. The remedy is *Calc. phos.*

Hysteria.—From the Greek, "uterus." It covers a variety of nervous affections. Schuessler says that *Kali phos.* "cures states of depression of the mind and of the body, hypochondriacal, ill-humor, neurasthenia, nervous insomnia and spasms caused by so-called irritable weakness." This covers hysteria. Other remedies may be called for if the hysteria is known to arise from a definitely known cause.

Impotence.—*Kali phos.*

Infants.—DIARRHŒA, COLIC.—Cholera infantum, in thin, weakly or scrofulous infants, *Calc. phos.* Cramp colic, *Magn. phos.* Diarrhœa, watery, passes undigested food, skin hot, *Ferrum phos.;* very foul, putrid, child in stupor, *Kali phos.;* sour, acid, raw, *Natrum phos.;* bilious, worse in wet weather, *Natrum sulph.;* with much perspiration, distended abdomen, *Silicea;* flatulent colic, *Natrum sulph.* CONSTIPATION.—Tongue white, what passes is light colored, *Kali mur.;* alternate with occasional attacks of diarrhœa (the best general laxative for infants), *Natrum phos.;* stool protrudes then goes back, *Silicea.* DENTITION, TEETHING.—Child is hot, feverish, restless, cross, *Ferr. phos.;* convulsions, *Magn. phos.;* weakly children, flabby, open fontanelles, scrofulous, *Calc. phos.;* the child sweats much, *Silicea.* GENERALITIES.—Fontanelles remain open too long, emaciation, bones soft, spongy, inclines to be bow-legged, abdomen too large, *Calc. phos.;* scrawny about the neck, *Natrum mur.;* head large and sweats much, *Silicea;* child seems to always

smell sour, *Natrum phos.* Feverish, cross,
fretful, *Ferrum phos.* Bear in mind the fact
that infants need a drink of pure, cool water
quite as often as adults require it, and that
they can suffer from thirst quite as severely
as adults.

Inflammation.—All inflammatory conditions require *Ferrum phos.* If septic suppuration sets in, *Kali phos.*

Influenza, Grippe.—This word comes
from "influence," which means "to control
by hidden power." Of late years it is common to call every attack of a bad cold
"grippe," which is synonymous with influenza, but this is an error. A cold comes on
gradually, but true grippe, or influenza,
strikes suddenly and is followed by a profound prostration. It is generally epidemic.
It is said to be caused by a bacillus, but until
we know what causes the bacillus, we are in
the same darkness concerning the disease as
ever. The remedy is *Natrum sulph.* If this
alone is used, there is, as a rule, prompt recovery and no bad after effects. If one has

used other remedies and recovered, but suffers from after effects, *Kali phos.* is the remedy.

Injuries.—Falls, contusions, blows, sprains and all manner of wounds require *Ferrum phos.*, for the fever or to prevent fever. When fever has subsided, *Kali mur.* will tend to prevent suppuration. If suppuration sets in, *Silicea.* If discharge becomes foul or gangrenous, *Kali phos.* If bones are broken, *Ferrum phos.* allays fever and *Calc. phos.* will greatly aid in knitting the bones.

Insomnia.—Inability to sleep is generally due to causes which must be treated to effect a cure. For insomnia without other causes, nervous insomnia, *Kali phos.*

Intermittent Fever.—Ague, chills and fever, paludal fever. The chief remedy is *Natrum sulph.* and it will generally cure all cases. If patient does not progress satisfactorily, give *Natrum mur.* In cases that have lasted for months or years, *Natrum mur.* 30x will often effect complete recovery.

Jaundice.—From the French, "jaune"—

yellow. *Natrum sulph.* will cure nearly every case. Should it fail, resort to *Kali mur.*

Kidney Affections.—In all kidney affections where there is inflammation, or discharge of blood, *Ferrum phos.* Albuminuria, *Kali sulph.* Albuminuria following scarlet fever, *Kali phos.* "*Silicea,*" writes Schuessler, "will prevent the formation of renal gravel."

Labor.—Parturition, child-birth. Where labor-pains are weak, ineffectual, tedious, *Kali phos.* Very painful, spasmodic, cramps, *Magn. phos.* After parturition, *Ferrum phos.*, to prevent fever.

Leucorrhœa.—From the Greek, "white" and "to flow." Milky white or grayish discharge, *Kali mur.* Yellowish, scalding, *Kali phos.* Greenish, *Kali sulph.* Watery, *Natrum mur.* Foul smelling, *Kali phos.*

Liver.—Patient turns yellow, *Natrum sulph.* Bilious, vomits, or bitter eructations, *Natrum sulph.* Pain on the right side, *Calc. sulph.* Tongue white, stools light colored, *Kali mur.* Abscess, *Silicea.* The chief biochemical remedy for the liver is *Natrum sulph.*

Locomotor Ataxia.—Tabes dorsalis. A nervous decay, affecting the walk. Said, from its nature, to be incurable. *Magn. phos., Kali mur.* or *Kali phos.* may give relief. It is a disease that comes on gradu-ally, often being years in developing.

Lumbago.—From the Latin for "loin." Remedy, *Ferrum phos.* Very sharp, light-ning-like pain, knife-like, *Magn. phos.*

Lupus.—From the Latin, "wolf." Eat-ing tetter. Remedy, *Kali sulph.* If this fails, *Natrum phos.* or *Silicea.*

Malaria.—(See "Intermittent.")

Marasmus.—(See "Atrophy.")

Mastitis.—From the Greek, "breast." Inflammation of the breast. The first rem-edy is *Natrum phos.*, which may produce re-absorption. If it goes on to suppuration, *Silicea.* For induration (hardening), *Calc. fluor.*

Measles.—Morbilli, or Rubeola. From the Danish, meaning "spot" or "speckle;"

eruptions. In the fever stage, *Ferrum phos.;* in secondary stage, cough, after effects, *Kali mur.* Rash recedes, suppressed, *Kali sulph.* If case becomes "watery," excessive secretion, *Natrum mur.* Albuminuria, *Kali sulph.*

Meningitis.—From the Greek, "membrane." Cerebro-spinal Meningitis, Spotted Fever. Inflammation of the membranes of the brain, or of spinal cord. The first remedy is *Ferrum phos.;* secondary, *Kali mur.*

Menstruation.—From the Latin for "month." The monthly flow. Painful, face flushed, vomits, *Ferrum phos.* Discharge late, dark, clotted, *Kali mur.* Menstrual colic, *Magn. phos.* Thin, watery, pale, *Natrum mur.* Debilitated, depressed, dejected, hysterical, *Kali phos.* Intense sexual desire, *Calc. phos.* Suppressed, by cold, *Ferrum phos.* With chilliness, *Silicea.* Offensive, *Calc. sulph.*

Mental.—Brain fag, insanity, general break-down, hypochondria, weak memory, nervous debility, *Kali phos.* Children undeveloped, idiocy, *Calc. phos.* Another remedy to be considered in brain affections is *Magn. phos.*

Morphine Habit.—*Natrum phos.* is said by some to be a "specific" for the habit and also *Kali phos.*, 6x. Which is best depends on the constitution of the patient. Probably *Kali phos.* will suit most cases.

Mumps.—From an Icelandic word meaning "to take into the mouth." Parotiditis; inflammation and swelling of the parotid gland. The remedy is *Kali mur.* If there is excessive salivation, give also *Natrum mur.* If there be high fever, *Ferrum phos.*

Neuralgia.—From the Greek, "nerve" and "pain." The chief remedy is *Magn. phos.*, though *Ferrum phos.* may be needed where there is fever, or the attack has been caused by a "cold" and there is inflammation.

Orchitis.—Inflammation of the testicles. The first remedy is *Ferrum phos.;* when inflammation has subsided, follow with *Kali mur.*, and if this does not completely cure, *Calc. phos.*

Ozæna.—From the Greek for "smell."
fetid smell. Ulceration of the nasal cavi-
ties with fetid discharge. Remedies, *Natrum phos.*, *Magn. phos.* or *Kali phos.* (See
also "Catarrh.")

Pains.—Pains felt on motion only, or
aggravated by it, *Ferrum phos.;* ameliorated
by motion, *Kali phos.;* worse at rest, *Calc.
phos.;* worse from warmth, better in cool air,
Kali sulph.; better from warmth, *Silicea;*
better by gentle pressure, *Magn. phos.*

Palpitation of the Heart.—With
fever, or congestion, *Ferrum phos.;* other-
wise, *Kali sulph.*

Paralysis.—From the Greek, "beside"
and "to loosen." A total, or partial, loss of
the power of motion in any part or parts.
The chief remedy is *Kali phos.*, though there
are other remedies for this condition, as
Natrum phos. for paralysis of lower extremities and *Magn. phos.* for paralysis agitans.
Probably the best treatment is to give *Kali
phos.* and *Magn. phos.*, alternately. Paresis

is an incomplete, or slight, paralysis; *Kali phos.* and *Magn. phos.* are the remedies.

Pemphigus.—From the Greek, "blister." Fever-blisters, water-blebs, vesicular. If the fluid is yellow, *Natrum sulph.;* if clear, *Natrum mur.*

Periostitis.—Formed from three Greek words, "around," "bone" and "inflammation." Inflammation of the periosteum, or immediate covering of the bone, the membrane. Remedies, *Silicea* and *Ferrum phos.*

Peritonitis.—From the Greek, "around" and "to stretch." Inflammation of the membrane of the abdomen. Remedy, *Ferrum phos.*, followed by *Kali mur.*

Phlebitis.—From two Greek words, "vein" and "inflammation." Inflammation of the veins, or of a vein. Remedies, *Ferrum phos.* and *Kali mur.*, alternately.

Phlegmasia Alba Dolens.—"Milk Leg." Remedies, *Natrum phos.* and *Kali mur.*, alternately; suppuration, *Silicea.*

Phthisis.—From the Greek, "to waste away." Consumption. Tuberculosis. At first threatening, patient anæmic, sweats, coughs and tires easily, *Natrum phos.*, later *Magn. phos.* When disease has developed, night sweats, fetid expectoration, *Silicea* and *Calc. sulph.* alternately. The best ally to these remedies is pure air, day and *night*, and light, open air exercise. (See "Scrofula.")

Pleuritis.—From the Greek, "rib" or "side." Pleurisy. Inflammation of the pleura. The remedy is *Ferrum phos.*, pain, stitch in the side, catching of the breath. The secondary remedy, for effusion, is *Kali mur.* In many cases *Ferr. phos.* is all that is required. Should pus form, *Calc. sulph.*

Pneumonia.—From the Greek, "lungs," and "to breathe." Congestion of the lungs. Pneumonitis. Inflammation of the lungs. The terse biochemic therapy for this disease given by Schuessler is: "*Ferrum phosphoricum.*" He then refers the reader to "exudations," *i. e.*, the secondary remedy must be chosen by the nature of the exudations; if fibrinous, *Kali mur.*; albuminous, *Calc. phos.*; if clear, watery, *Natrum mur.*;

if yellowish, watery, *Natrum sulph.;* if fetid, *Kali phos.;* if thick, yellow pus, *Kali sulph.,* and should indurations, hardening, remain, *Calc. fluor.*

Polypus.—From the Greek, "many" and "foot" or "feet." Tumors on mucous membranes, often pear-shaped, fig-warts. Remedy, *Calc. phos.*

Pruritus.—Itching. For itching of knees, elbows, genitals or anus, *Magn. phos.;* itching eruptions, *Kali sulph.* Nettle rash, *Calc. phos.*

Psoriasis.—From the Greek "to scratch" or "rub." Scaly tetter. Dry, scaly skin. Tetter. Remedy, *Kali sulph.*

Puerperal Fever. — Child-bed fever. Remedies, *Ferrum phos.* and *Kali mur.,* alternately. With mania, *Kali phos.*

Rachitis.—From the Greek, "the spine." Rickets. A disease leading to curvature of the spine. The remedy is *Calc. phos.* If fetid diarrhœa supervene, alternate *Kali phos.,* or if there be acidity, *Natrum phos.*

Rheumatism.—An indefinite word from

the Greek, "to flow," covering a variety of ailments, muscular, articular, synovial, cardiac, cerebral, etc., etc. The chief remedy for rheumatism in general, of shoulders, joints, back, arms, etc., is *Ferrum phos.*, with *Kali mur.* as a secondary remedy after pain has subsided, or in alternation. For what is known as inflammatory rheumatism, the remedy is *Ferrum phos.* alternated with *Natrum phos.* Chronic rheumatic conditions, *Kali phos.* Shifting pains, worse in a warm room, *Kali sulph.*

Scarlet Fever.—Scarlatina. The primary remedy is *Ferrum phos.* After rash has subsided, *Kali mur.* For discharges from the ear that follow in some cases, *Kali sulph.* Post-scarlatinal dropsy, *Natrum mur.*, or if this fails to give relief, *Kali phos.*

Sciatica.—Neuralgia of the sciatic nerve. Ischias. If brought on by cold or exposure, *Ferrum phos.*; if pain is exceedingly sharp, lightning-like, *Magn. phos.*; for dull, steady pain, *Kali phos.*

Scrofula.—From the Latin, for "sow,"

"swine." When the glands are swollen, or as a primary remedy, *Natrum phos.*, but at an advanced stage, when glands break down, *Magn. phos.* Schuessler holds that scrofula and tuberculosis are in origin the same and that these two remedies are the essentials, though other biochemical remedies are needed for "the catarrhal symptoms and the hæmorrhages from the lungs." (See "Phthisis.")

Scurvy.—Scorbutus. Scurf. Remedy, *Kali phos.*

Sea-Sickness.—*Kali phos.* and *Natrum phos.*, alternately.

Septicæmia.—From the Greek, for "putrid." Blood poisoning. *Kali phos.*

Skin Diseases.—These are generally manifestations of some constitutional malady and the remedies should be as far as possible for their cause—the internal malady. The following, in general, are the remedies for the skin. For heat congestion, pimples not suppurating, *Ferrum phos.* Mealy scurf, vesicles with fibrinous contents, *Kali mur.*; yellowish, *Calc. phos.;* white scales, or clear vesicles, *Natrum mur.;* greenish, *Natrum sulph.;* pus, *Silicea;* greasy crusts, or ichorous, oozing or tinged with blood, *Kali phos.;* hard, horny, growths, or crusts, cracked, *Calc. fluor.;* nettle rash, *Kali phos.;* barber's itch, *Magn. phos.*

Small-Pox.—Variola. This disease may be distinguished by the fact that it generally begins with chill, fever, headache and backache. About the fourth day papules appear that *feel like shot beneath the skin;* later these break and are the "pocks," as the old form of spelling has it. *Kali mur.* is the chief remedy, though if fever be especially marked, *Ferrum phos.* may be alternated during fever. When pustules break, *Natrum phos.* Should the case become "putrid," *Kali phos.*

Sore Throat. — Throat dry and red, *Ferrum phos.;* swollen, grayish patches, swallowing is painful, hawking of phlegm, *Kali mur.* "Clergymen's sore throat," *Ferrum phos.* Throat ulcerated, bad smell, *Kali phos.*

Spasms.—Convulsions. In teething children, *Ferrum phos.* Drawn, twitching, rigid, tonic spasms, "fits," writer's cramp, *Magn. phos.* Night convulsions, *Calc. phos.*

Spermatorrhœa.—Nightly emissions, *Natrum phos.* For extreme debility resulting from the habit, broken down, neurasthenic, *Kali phos.*

Spinal Diseases.—Spinal anæmia, softening of the spinal cord, *Kali phos.* Pains in the spine, diseases dating from injuries to the spine, *Magn. phos.* Spinal anæmia, curvature, abscesses, *Calc. phos.* Spinal irritation, *Natrum mur.*

Sprains.—*Ferrum phos.* and *Calc. fluor.*, alternately.

Stomach.—Inflammation, pain, gastralgia, vomiting, *Ferrum phos.* Cramps, *Magn. phos.* Full, uncomfortable feeling, tongue yellow, *Kali phos.* Umbilical colic, flatulent colic, must bend forward, *Magn. phos.* Acid, sour, heart-burn, *Natrum phos.* Ulcerations of the stomach, *Kali phos.* Enlargement of the stomach, *Kali phos.*

Styes.—Tumor on edge of eye-lid, *Silicea* and *Calc. fluor.*, alternately.

Sunstroke.—*Natrum mur.* and *Kali phos.*, alternately.

Syphilis.—From the Greek, "hog" and "loving," a term originating among primitive people. Soft chancre, *Kali mur.* Phagedenic, "eating," chancre, *Kali phos.* Hard chancre, *Calc. fluor.*

Teething.—To aid and assist the teething process, *Calc. phos.* If there is fever and spasms, *Ferrum phos.* If there are spasms but no fever, *Magn. phos.* Slavering, *Natrum mur.*

Tetanus.—Lockjaw, *Magn. phos.*

Tongue.—White, gray, *Kali mur.*; brownish-green, *Natrum sulph.*; yellow, moist, *Natrum phos.*; yellow, mucous, *Kali sulph.*; as though covered with mustard, *Kali phos.*; clean, moist, minute bubbles, *Natrum mur.*; red, inflamed, *Ferrum phos.*

Tonsillitis.—Amygdalitis. Inflammation of the tonsils; suppurative tonsillitis, quinsy. Inflamed, red, swollen, pain on swallowing, *Ferrum phos.* and *Kali mur.*, alternately. When suppuration discharges, *Calc. sulph.*

Toothache.—Odontalgia. From cold, ininflamed, *Ferrum phos.* This remedy will meet most cases. Neuralgic, *Magn. phos.* In decayed teeth, *Silicea.*

Tuberculosis.—(See "Scrofula" and "Phthisis.") When disease is suspected, *Natrum phos.*, "but caseous degeneration requires *Magn. phos.*" This disease is generally developed by bad air, confining work and mal-nutrition, and unless these can be corrected, medicine can but give temporary relief.

Tumors.—Hard, or firm, shiny, swellings, *Calc. fluor.* Tumors or growths on very young children, polypi, nasal polypi, *Calc. phos.* Epithelioma, "skin cancers," or tumors, *Kali sulph.*

Typhoid.—From the Greek, "smoke," involving the idea of stupor and fever. Typhus. The remedy is *Kali phos.*, with *Natrum mur.* where the case falls into profound stupor. Let the patient have all the water (cold and pure) wanted, and do not press any food on him. Many cases are lost by over-feeding, especially during convalesence, when patient is apt to be hungry. Typhus, or "jail fever," is the same as typhoid, but in a more malignant form; indeed, typhoid is, literally, "resembling typhus."

Typhlitis.—Inflammation of the cæcum, practically the same as appendicitis. *Kali mur.* and *Ferrum phos.*, alternately. "We have no remedy which is the peer of *Ferrum phos.* as a fever remedy, whether idiopathic or symptomatic, and none better than *Kali mur.* to cause the absorption of infiltration." Dr. I. E. Nicholson.

Ulcers.—If there be fever, *Ferrum phos.* and *Kali mur.*, alternately. Thin, ichorous discharge, *Silicea.* Exuding yellow, creamy pus, *Natrum phos.* Foul smelling, *Kali phos.* Varicose, *Calc. fluor.*

Urinary Disorders. — Incontinence, dribbling, enuresis or wetting the bed; or

cannot pass water, retention; or it passes
hot and scalding, *Ferrum phos.* Also, for
inability to pass urine, *Calc. phos.* may be
needed, and, if spasmodic, *Magn. phos.*
Sandy deposits, *Natrum sulph.* Much pale
urine, polyuria, *Natrum mur.* Urine foul,
Kali phos.

Vaccination.—For ill effects following the
operation and for the running sores resulting,
Silicea.

Vertigo.—From the Latin, "to turn."
If caused by a rush of blood, *Ferrum phos.*
If nervous, *Kali phos.* In old people, *Silicea.*
When rising, *Natrum mur.* From biliousness,
Natrum sulph. With vomiting, *Natrum phos.*

Vomiting.—Of bile, *Natrum sulph.;* of
clear, thready mucus, *Natrum mur.;* of white
mucus, *Kali mur.;* of red blood, *Ferrum phos.;*
of dark blood, *Kali phos.;* sour, *Natrum phos.;*
of food soon after eating, *Ferrum phos.;* of
children during teething, *Calc. fluor.*

Whooping Cough.—Pertussis. At development of disease, *Ferrum phos.* and *Kali
mur.*, alternately; should these fail to check

the disease and it develops into the characteristic, nervous "whoop," *Magn. phos.*
Should case go on to extreme prostration,
with death imminent, *Kali sulph.*

Worms.—As a general vermifuge, *Natrum
phos.* 2x. For ascaris lumbricoides, long
intestinal worms, *Natrum mur.* 12x.

Yellow Fever.—For this and kindred
tropical fevers, *Natrum sulph.;* for black
vomit stage, *Kali phos. Ferrum phos.* is
called for during preliminary fever and may
be used intercurrently.

CPSIA information can be obtained at www.ICGtesting.com
Printed in the USA
LVOW05s1753140414

381651LV00030B/1041/P